THE BEE BOOK

DK

WHAT'S THE **BUZZ** ABOUT **BEES**?

Bees have been on planet Earth for 100 million years. That's a lot longer than us humans and even longer than some dinosaurs! They are insects that are important for our planet. This is because of the special relationship between bees and plants. But many bees are disappearing.

Let's find out more about the world of bees, and why there are fewer buzzing about...

THERE ARE AROUND 20,000 DIFFERENT SPECIES OF BEES

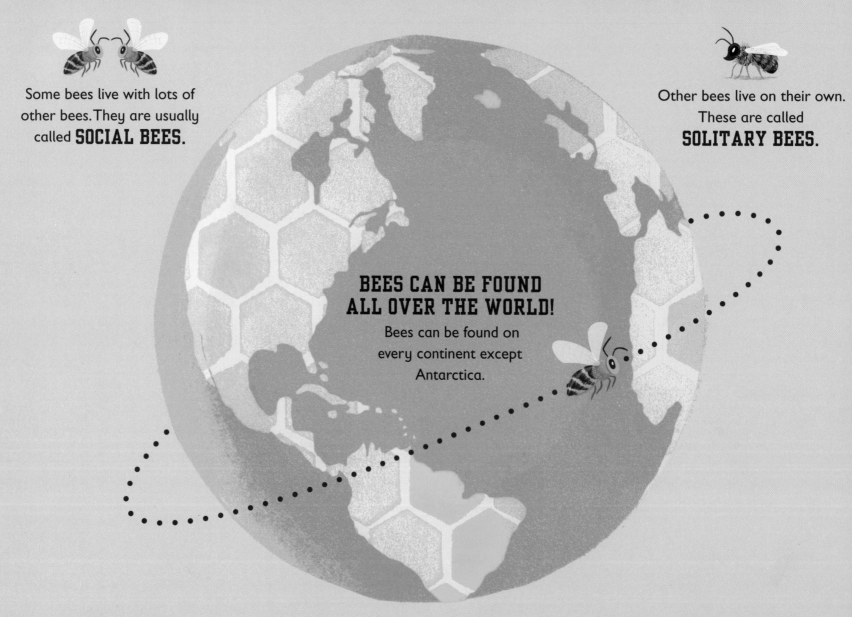

Some bees live with lots of other bees. They are usually called **SOCIAL BEES.**

Other bees live on their own. These are called **SOLITARY BEES.**

BEES CAN BE FOUND ALL OVER THE WORLD!

Bees can be found on every continent except Antarctica.

SOME BEES ARE VERY BIG...

The **WALLACE'S GIANT BEE** is the largest bee in the world at almost 4cm (1½in) long. It can be found in Indonesia, and lives in termite mounds.

SOME BEES ARE VERY SMALL...

The **DWARF BEE** is the world's smallest bee at just 2mm (⅛in) long. That's 20 times smaller than the Wallace's Giant Bee. You might need a magnifying glass to see this one!

SOME BEES ARE ANGRY...

KILLER BEES escaped from a science experiment and have since spread across South and Central America. These bees get very defensive, stinging ten times more often than other bees.

SOME BEES CAN COOK...

When **JAPANESE HONEY BEES** spot an enemy hornet, they form a hot bee ball by surrounding the dangerous suspect, and they cook it with their body heat.

SOME BEES YOU MIGHT KNOW ALREADY...

BUMBLE BEES are the fluffy bees that you may have seen buzzing around your garden. But you might not know that the Old English word for bumble bee is "dumbledore"!

BUT THE MOST FAMOUS BEE IS THE HONEY BEE...

WHAT IS A HONEY BEE?

Like all bees, a honey bee is an **INSECT**. All insects have six legs, and generally have one or two pairs of wings.

Honey bees live together in a large group called a **COLONY**. Every colony contains a queen bee and worker bees.

The colony lives in a **NEST**. Some nests are wild, and some are kept by humans in beehives.

FROM FLOWERS A HONEY BEE COLLECTS...

POLLEN

NECTAR

IN THE BEEHIVE A HONEY BEE MAKES...

BEESWAX

HONEY

POLLEN is a powder made by flowers. Honey bees collect pollen to feed to baby bees.

NECTAR is a sweet, sticky liquid made by flowers. The honey bees love it! They use nectar to make honey.

BEESWAX is used by honey bees to make hexagonal cells, which together make a honeycomb. Honey bees fill the comb cells with honey, pollen, and eggs.

HONEY is made from the collected nectar. Honey bees store honey in the comb cells and eat it when there are no flowers available.

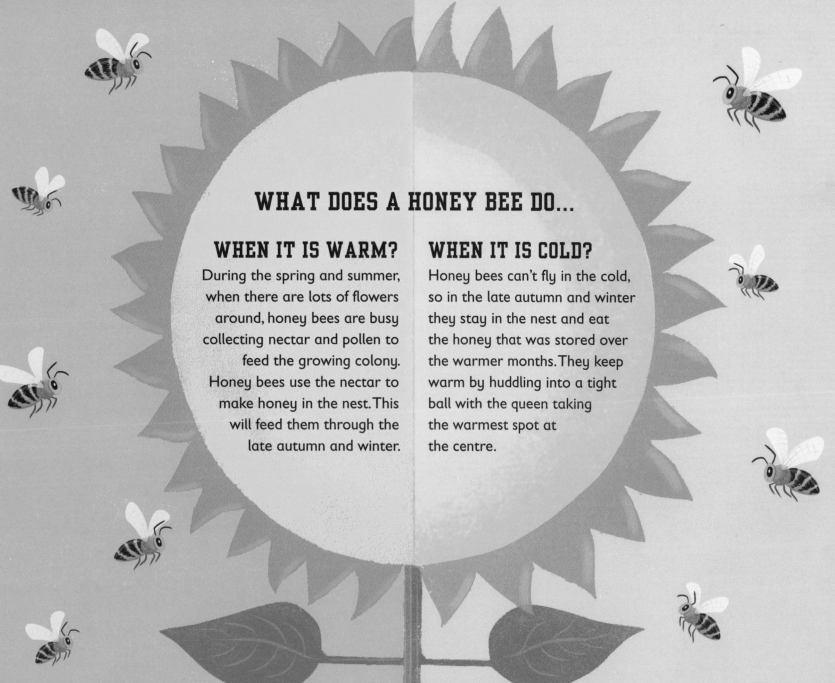

WHAT DOES A HONEY BEE DO...

WHEN IT IS WARM?

During the spring and summer, when there are lots of flowers around, honey bees are busy collecting nectar and pollen to feed the growing colony. Honey bees use the nectar to make honey in the nest. This will feed them through the late autumn and winter.

WHEN IT IS COLD?

Honey bees can't fly in the cold, so in the late autumn and winter they stay in the nest and eat the honey that was stored over the warmer months. They keep warm by huddling into a tight ball with the queen taking the warmest spot at the centre.

THE BEE'S KNEES

A honey bee's body is made up of three different sections. It has a **HEAD**, a middle section called the **THORAX**, and the end section which is the **ABDOMEN**. Its body parts make it very skilled at flying, finding flowers, and collecting nectar and pollen.

LEGS carry pollen to the nest. The front legs push pollen off the bee's body onto its back legs where it is packed into **POLLEN BASKETS.**

A honey bee doesn't have a nose. Instead it uses its **ANTENNAE** to smell as well as taste.

A honey bee has **FIVE EYES!** It has two big eyes, which are excellent for detecting flowers, and three small eyes that detect light.

The **MOUTH** has a long tongue which the honey bee uses like a straw to reach into flowers and suck up lots of nectar.

A honey bee stores nectar in a **HONEY STOMACH** while it is carrying it to the nest.

SPIRACLES are tiny holes all over a bee's body that the bee uses to breathe through.

WHY DO BEES BUZZ?

You have probably heard the buzz of a honey bee, but have you ever thought about what causes it to make that sound? For a bee to fly, its wings have to beat very fast. This makes vibrations in the air that we hear as a buzz sound. For some bees, buzzing is very important. The **bumble bee** lets out a loud buzz whilst shaking its body and wings. This shakes pollen off flowers, allowing the bumble bee to collect more pollen to take back to the nest.

THE STINGER

Honey bees use their stingers to protect the colony.

DO ALL HONEY BEES STING?

No, only the female honey bees can sting. The male bees are harmless.

DOES A HONEY BEE DIE WHEN IT STINGS?

Yes, when it stings a mammal (like us!) a honey bee dies. Do you see how the stinger has a jagged surface? When a honey bee stings, the jagged spikes get stuck into the mammal's skin and the stinger is pulled off the bee. A honey bee can't survive without a stinger. The queen, however, is lucky. Her stinger has a smooth surface, allowing her to sting, sting, and sting again without doing her any harm at all!

WE LOVE **HONEY BEES** AND WE LOVE **HONEY**

Honey is really important to honey bees because they need it to survive the cold months. In fact, honey is so important that worker honey bees spend most of their lives making it. With the hard work of honey bees comes a sweet reward! Do you like to eat honey as much as honey bees do?

Let's find out how honey bees make honey...

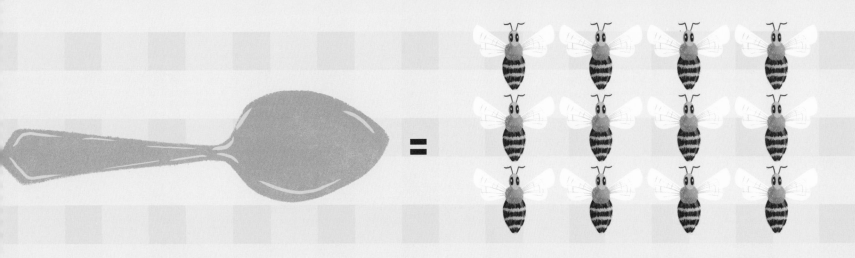

ONE TEASPOON OF HONEY

THE LIFETIME'S WORK OF 12 HONEY BEES

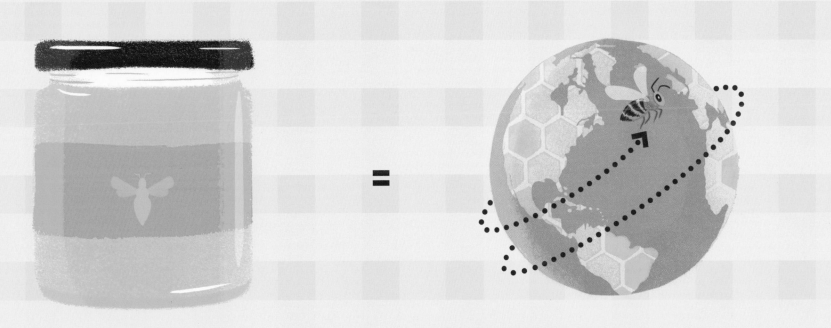

ONE JAR OF HONEY

88,500 KM (55,000 MILES) OF FLYING
(That's one and a half times around the world!)

WHERE DOES HONEY COME FROM?

Some honey bees live and make honey in the wild. Others are domestic and make honey that is collected by humans. These honey bees are kept in beehives and are cared for by **BEEKEEPERS**. It is this special relationship between honey bees and beekeepers that allows us to find honey in the shops and on our breakfast tables.

In the wild, honey bees make nests in rocks, hollow trees, and caves.

Beekeepers keep their honey bees in **BEEHIVES**. These homes protect honey bees from the cold and rain, as well as keeping them cool in the summer. Inside the hives, honey bees build honeycombs and fill them with honey that can be collected by beekeepers and put into jars.

Humans have been collecting honey for 13,000 years — that's a very long time! Jars of honey were even found in Egyptian tombs. The Pharaoh Tutankhamun loved honey so much that he was buried with it!

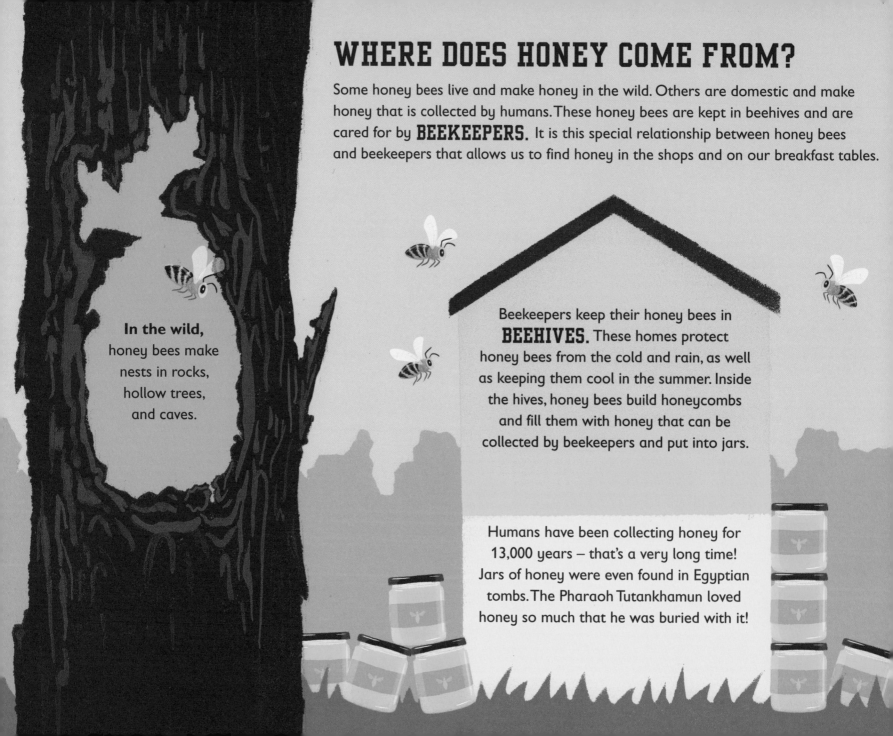

MEET THE BEEKEEPERS
WHY DON'T THEY GET STUNG?

BECAUSE THEY KNOW THE HONEY BEE

Knowing about how honey bees behave and what they get up to inside the hive is important for keeping a healthy colony. Having this understanding means that a beekeeper can stay calm when checking the hive and collecting honey. A calm beekeeper means that the bees will be calm too.

BECAUSE THEY USE A SMOKER

A smoker pumps out smoke around the nest and calms the honey bees. The smell of smoke makes the bees think that there is risk of a fire, and so they start to eat honey in case they need to abandon the hive. Stinging is the last thing on the honey bees' minds if they think there might be a fire!

BECAUSE THEY WEAR A BEE SUIT

A bee suit has gloves, a hood, and a veil. Every part of a beekeeper needs to be covered so that they can't get stung.

WHAT'S HAPPENING IN THE HIVE?

There are about **35,000 HONEY BEES** in a beehive. The colony has three types of bees: a queen, worker bees, and drones. Each type relies on one another to keep the colony healthy.

THE QUEEN
[FEMALE]

HOW MANY: There is just one queen in a beehive.

HER LIFE SPAN: 5 years.

HER JOB: As the mother to most of the bees in a hive, the queen must lay eggs. She can lay 2,000 eggs every day!

THE WORKER
[FEMALE]

HOW MANY: Thousands.

HER LIFE SPAN: 40 days.

HER JOB: The worker bee has the most jobs to do including cleaning and guarding the hive, foraging, and making honey.

THE DRONE
[MALE]

HOW MANY: Hundreds during the summer.

HIS LIFE SPAN: Just a few weeks.

HIS JOB: A drone bee doesn't do much work, but is needed in the summer to mate with a queen from another colony. In autumn the drones are pushed out of the hive.

HOW DO HONEY BEES MAKE HONEY?

1.

Honey bees **collect** nectar from flowers with their tongues and store it in their honey stomachs.

2.

These bees fly back to the hive and **pass** the nectar to another bee using their mouths.

3.

This bee makes spit bubbles with the nectar until it turns into honey, which is then **stored** in wax cells.

The warm hive is the perfect place to make honey. Bees flap their wings to keep air flowing through the hive, keeping the colony dry.

4.

The bees **seal** the honey in each cell with a wax lid on top to keep it fresh.

WHAT MAKES BEES SO BEE-RILLIANT?

Not all bees make honey like the honey bee does, but all bees do play an important part in feeding us much of the food that we eat. This is because bees pollinate plants. Some plants need pollination to make new plants, and to grow seeds and fruit that we eat.

Let's find out how bees pollinate plants...

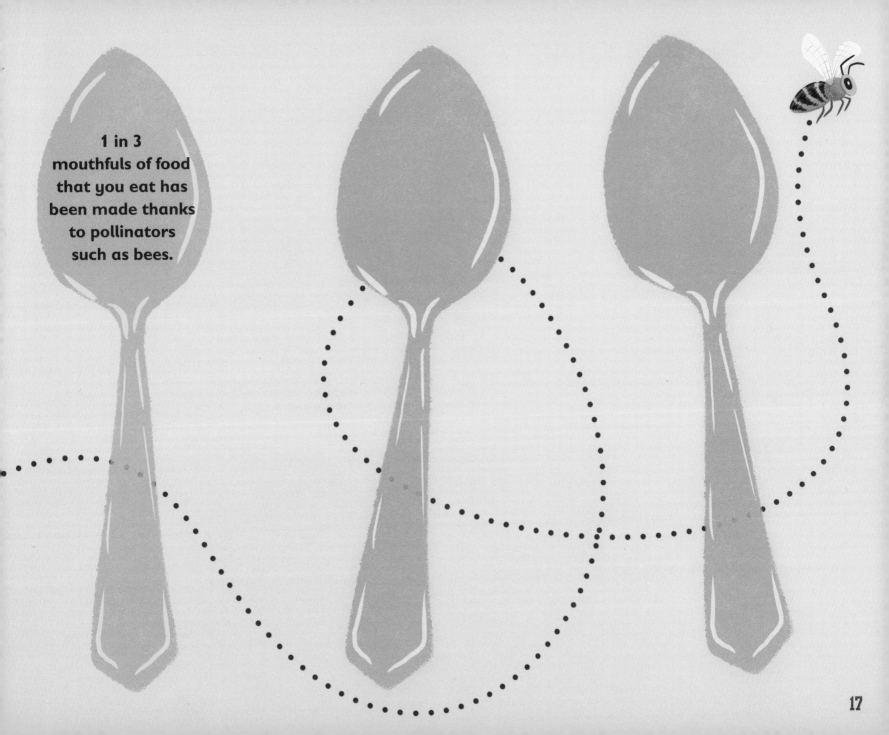

1 in 3 mouthfuls of food that you eat has been made thanks to pollinators such as bees.

WHAT IS POLLINATION?

The process of pollen moving from one flower to another is called **POLLINATION**. Flowers need pollination to make seeds. **REPRODUCTION** is when these seeds grow into new plants. But the flowers need help moving pollen. Sometimes this help comes from the wind, and sometimes it comes from animals, such as bees. Bees are **POLLINATORS** because they move pollen from flower to flower.

HOW DO BEES POLLINATE PLANTS?

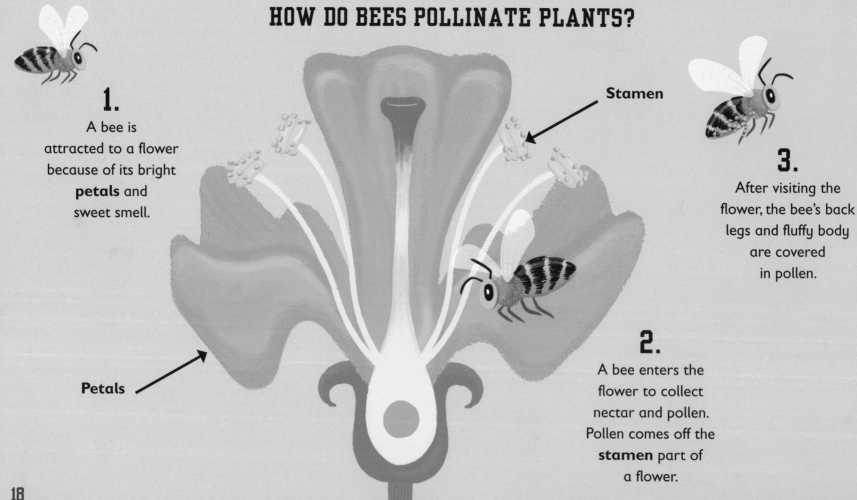

1.
A bee is attracted to a flower because of its bright **petals** and sweet smell.

Stamen

3.
After visiting the flower, the bee's back legs and fluffy body are covered in pollen.

Petals

2.
A bee enters the flower to collect nectar and pollen. Pollen comes off the **stamen** part of a flower.

4.

The bee visits another flower to collect even more pollen and nectar.

Stigma

5.

At the next flower, the pollen falls off the bee's body and gets stuck on the sticky part of the flower, which is called the **stigma**.

6.

The pollen enters the flower at the stigma and grows a tube to the ovary.

Here, it meets the egg.

After the pollen has joined with the egg, the egg starts to become a **seed**.

The seed can then grow into a new **plant**.

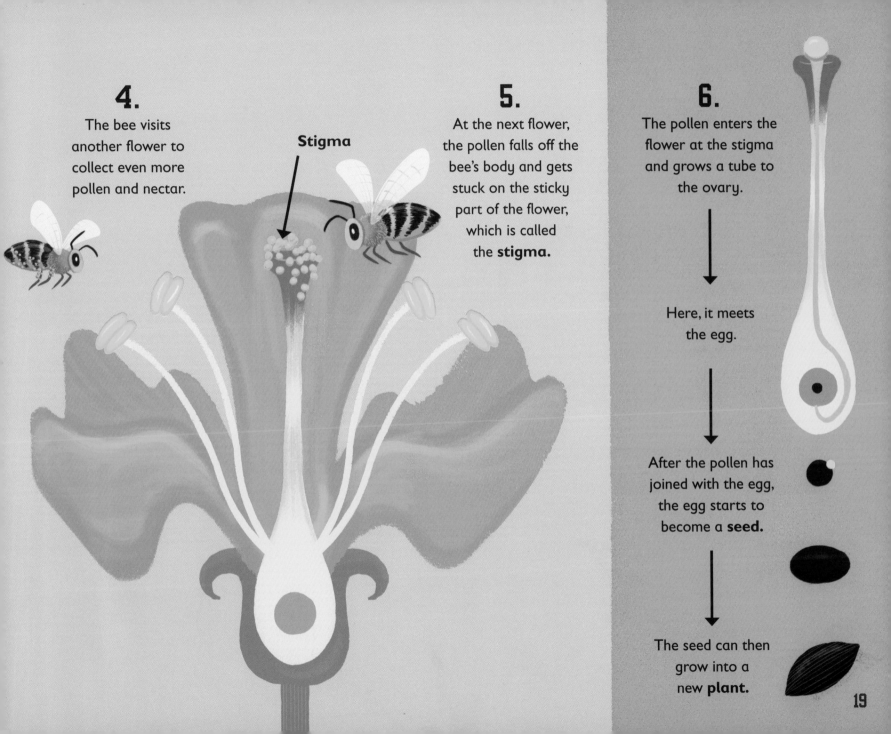

WHY DO WE NEED POLLINATION?

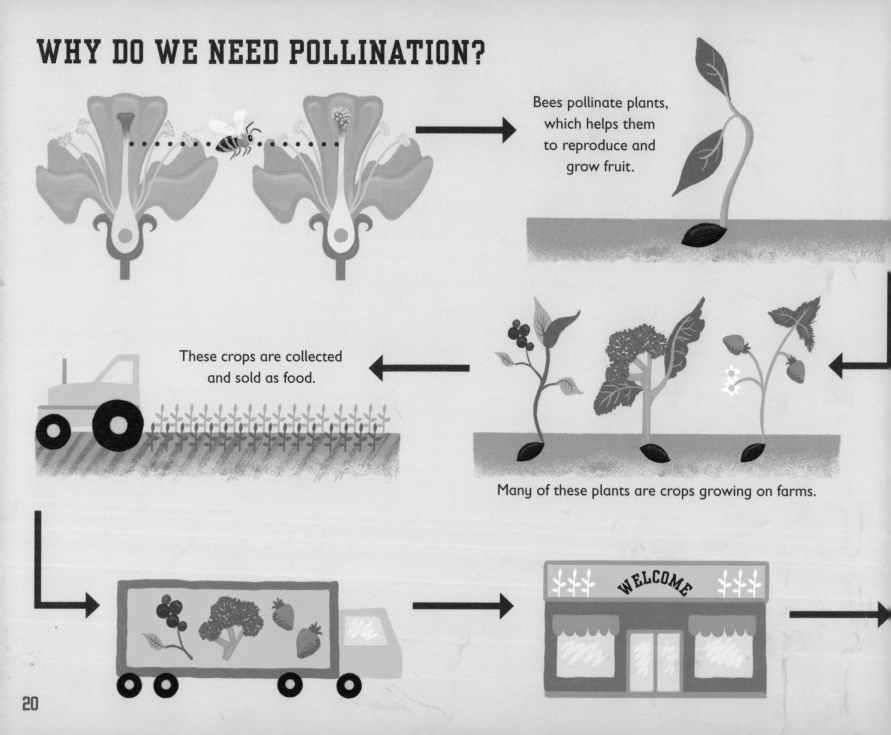

Bees pollinate plants, which helps them to reproduce and grow fruit.

These crops are collected and sold as food.

Many of these plants are crops growing on farms.

WELCOME

Bees pollinate almond trees. Almonds are used to make **cereal.**

Bees pollinate the coffee plant. This plant grows coffee beans, which are used to make **coffee.**

Bees pollinate orange trees. These trees grow the fruit used to make **orange juice.**

Bees pollinate the strawberry plant. Strawberries are used to make **jam.**

Breakfast is served! One third of the food that we eat comes from crops pollinated by animals.

AND THERE'S A LOT MORE TO KNOW ABOUT THE **HONEY BEE**

Honey bees are very busy making honey and pollinating plants. But there's a lot more that goes on in a honey bee's world. What does the queen bee do? How is a honey bee born? Can honey bees talk to each other?

Let's discover more about the life of a honey bee...

HOW IS A WORKER HONEY BEE BORN?

DAY 0-3

The queen bee lays an **EGG.**

All honey bees start off as an **egg**. The **queen** lays the eggs in **comb cells** that have been made by the worker bees. This gives each egg its own space to grow into a bee.

DAY 4-10

The egg hatches into a **LARVA.**

A **larva** looks like a tiny, white worm. Like a baby, a larva needs lots of food for it to grow. **Nurse bees** give it all the food it needs.

DAY 11-20

The larva spins a cocoon, becoming a **PUPA.**

When a larva has grown big enough, it spins a silk cocoon for protection whilst it becomes a pupa. A pupa starts to grow the body parts it needs to become an adult bee.

DAY 21

The pupa grows into an **ADULT BEE.**

When fully grown, the **adult bee** bites its way out of the **comb cell**. It is ready to start its busy life.

THE LIFE OF A WORKER BEE

A female worker bee has lots of different jobs to do. She will be a cleaner, a nurse, and a builder all in one lifetime!

The worker bee **cleans out the comb cell** where she was born.

DAY 1-2 CLEANING

The worker **builds comb cells** out of beeswax, and helps to fill them with honey.

DAY 12–17 BUILDING

DAY 18–21 GUARDING

The worker looks out for danger and **guards the nest's entrance.**

DAY 3–11 NURSING

As a nurse bee, the worker has the job of **feeding larvae.**

DAY 22+ FORAGING

Nurse bees feed larvae royal jelly, pollen, and honey.

If a larva is fed royal jelly for more than three days, it will grow into a queen bee.

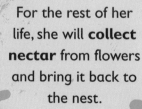

For the rest of her life, she will **collect nectar** from flowers and bring it back to the nest.

25

WHO IS QUEEN BEE?

Even though we call her the queen, she isn't really in charge of all the bees in the nest. It is better to think of the queen bee as more of a super-mum. This is because she is the mother of most of the bees in her nest. It is the queen's job to populate the colony by laying eggs that will hatch into the next generation of bees.

WHAT MAKES A HONEY BEE A QUEEN?

The queen has a **LONGER** and more pointed body than the other bees in the colony.

Like a celebrity, the queen always has other worker bees surrounding her. These **ATTENDANTS** feed her, groom her, and pass on her scent to the other bees in the nest.

The queen has a special scent that the bees in the colony can detect. Her **QUEEN SCENT** tells the colony that she is alive and healthy. This message is important, as without a queen, a colony can't survive.

The queen can **CONTROL** whether an egg will hatch into a male drone bee or a female worker bee.

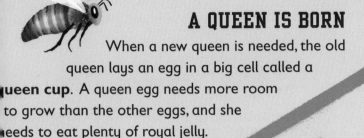

A QUEEN IS BORN

When a new queen is needed, the old queen lays an egg in a big cell called a **queen cup**. A queen egg needs more room to grow than the other eggs, and she needs to eat plenty of royal jelly.

THE FIGHT TO BE QUEEN

There can be only one queen in a nest, so if more than one queen is born, they fight by stinging each other. The survivor is crowned the queen.

WHAT DOES THE QUEEN DO?

EGG LAYING

The queen spends her life laying eggs. She can lay as many as one egg every twenty seconds! If she slows down at laying eggs then the colony might replace her with a new queen.

MEETING A DRONE

The queen takes a **mating flight** to meet a drone bee. She flies to where thousands of drone bees have gathered to meet her. When she returns to the nest she is ready to start laying eggs.

Honey bees need to be able to tell each other where to find the best flowers. But honey bees can't talk. Instead, they dance! Bees communicate inside the hive using the **WAGGLE DANCE**.

When a honey bee does the waggle dance, she walks around in two loops and shakes her body. The angle of the dance tells other bees the **DIRECTION** of the flowers.

The dance also tells other bees **HOW FAR** away the flowers are. The longer a bee dances for, the further away the flowers are from the hive. A short waggle dance means that the flowers are close.

Honey bees carefully watch the waggle dancer's every move. They even notice the dancer's smell! They smell flowers to find the same **TYPE OF FLOWER** that the dancer visited.

WHAT IS A SWARM?

A large group of honey bees flying together is called a swarm. A swarm forms when the queen, along with some of the colony, leaves the old nest to find a new nest.

A swarm is made up of one queen and up to 20,000 worker bees. **Can you spot the queen in the swarm?**

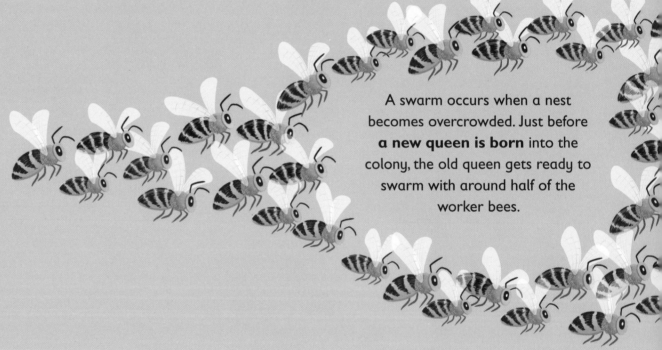

A swarm occurs when a nest becomes overcrowded. Just before **a new queen is born** into the colony, the old queen gets ready to swarm with around half of the worker bees.

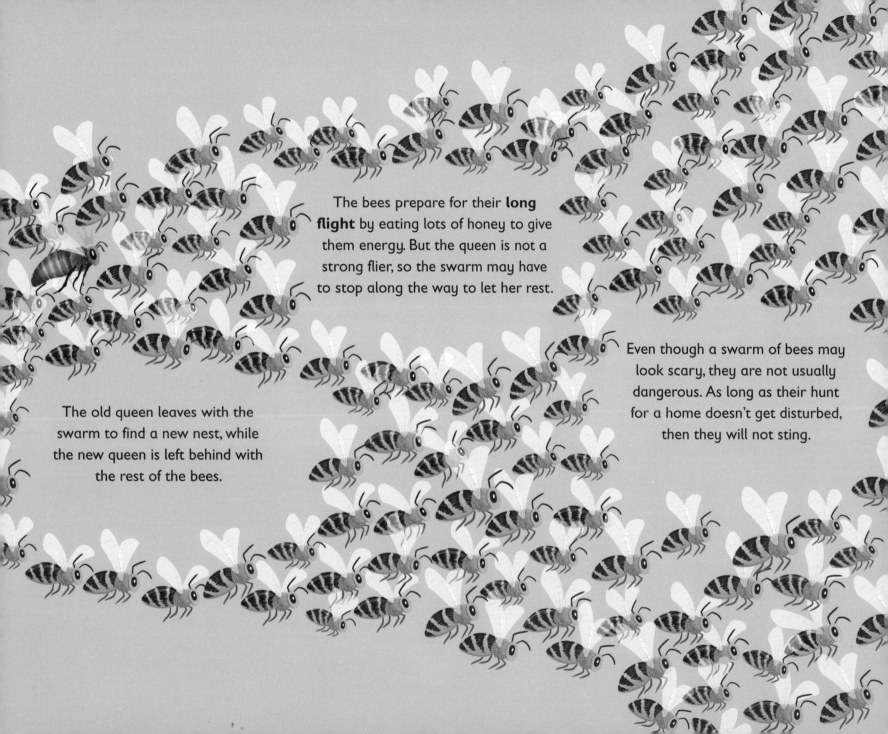

The bees prepare for their **long flight** by eating lots of honey to give them energy. But the queen is not a strong flier, so the swarm may have to stop along the way to let her rest.

The old queen leaves with the swarm to find a new nest, while the new queen is left behind with the rest of the bees.

Even though a swarm of bees may look scary, they are not usually dangerous. As long as their hunt for a home doesn't get disturbed, then they will not sting.

WHO EATS HONEY BEES?

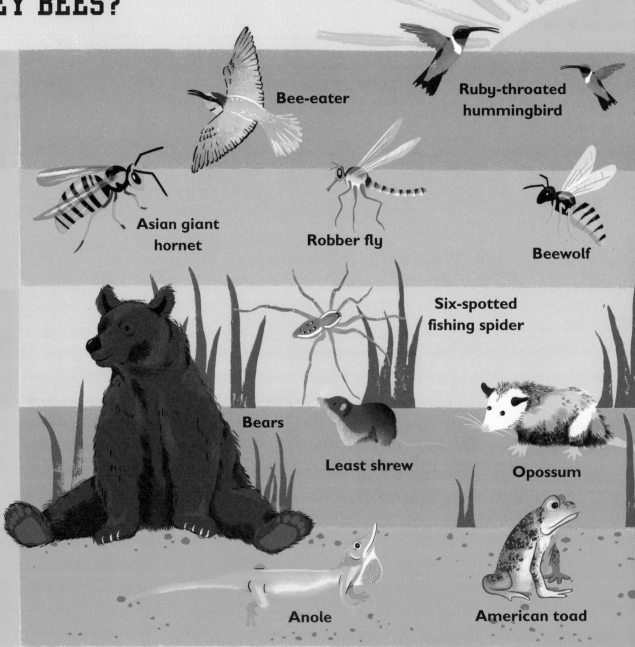

BIRDS

Bee-eater

Ruby-throated hummingbird

INSECTS

Asian giant hornet

Robber fly

Beewolf

SPIDERS

Six-spotted fishing spider

MAMMALS

Bears

Least shrew

Opossum

REPTILES AND AMPHIBIANS

Anole

American toad

These animals eat honey bees, and some, like the honey badger, love the sweet taste of honey.

Common grackle

Tyrant flycatcher

Dragonfly

Chinese mantis

Bald-faced hornet

Green lynx spider

Golden rod spider

Raccoon

Honey badger

Skunk

Bull frog

Wood frog

SO WHAT'S THE STING IN THE TAIL?

Bees may be little insects, but they are very important to our environment. They have an effect on plants and animals all over the world, from a small flower to a big, hairy bear. But many types of bees are disappearing, and it's not because animals are eating them. There is a much bigger problem that's making it difficult for bees to survive.

What's behind the problem of bee decline?

WHY ARE THERE FEWER BEES ABOUT?

There is not just one cause. There are lots of reasons why many types of bees are in decline. When we think about the impact of these **FACTORS** put together, we can start to see why bees are in trouble.

THE FACTORS:

1. CLIMATE CHANGE

Climate change can change the environment where bees and plants live (by making it hotter, for example). This makes it difficult for plants and bees to survive in the place where they are adapted to live.

2. LOSS OF HABITAT

A bee's habitat is the place where it lives. Bees like to nest in areas where there are all sorts of wildflowers and grasses growing. But humans are destroying these habitats to build more roads, cities, and farmland. This means that there are fewer places for bees to live.

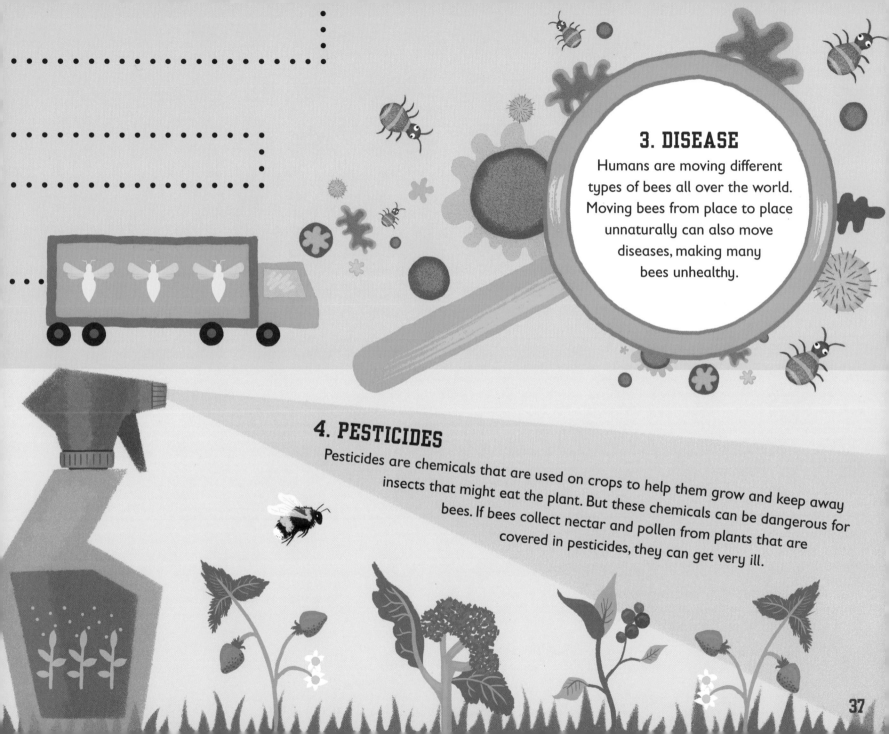

3. DISEASE

Humans are moving different types of bees all over the world. Moving bees from place to place unnaturally can also move diseases, making many bees unhealthy.

4. PESTICIDES

Pesticides are chemicals that are used on crops to help them grow and keep away insects that might eat the plant. But these chemicals can be dangerous for bees. If bees collect nectar and pollen from plants that are covered in pesticides, they can get very ill.

WHAT WOULD WE DO WITHOUT BEES?

THE CUPBOARDS WOULD BE EMPTY

Without bees we would risk losing some types of fruit and vegetables.

We would risk losing many meat and dairy products too, because we feed farm animals with food that is made from pollinated plants.

WHAT WOULD WE HAVE FOR BREAKFAST?

Without a variety of fruit and vegetables to choose from, our diet would be limited and unhealthy.

We would probably eat a lot more bread, as bread can be produced without bees.

WE WOULD HAVE TO FIND NEW WAYS TO POLLINATE

Can you imagine having to do the job of a busy bee? In areas of China where there are few bees and other insect pollinators, people are pollinating their apple trees by hand! Workers climb up apple trees, carrying small pots of pollen and brushes, which they use to pollinate each flower on the tree.

Could robots do the job of a bee? Scientists are working on a **ROBOBEE** that will have the technology to pollinate. But we could never really replace bees. Robots and human workers are both very expensive ways to do something that nature does for free.

WHY DO WE NEED TO HELP THE BEES?

Not only are bees important to the production of food through pollination, they also help support a huge variety of wildlife. With fewer bees, other plants and animals in the environment could be in trouble too.

Buzzing **BEES** spend their days pollinating plants.

These plants include **CROPS**, such as fruit and vegetables.

With fewer bees, there could be fewer fruit and vegetables. **FOOD** could become more expensive.

By pollinating plants, the bees help **WILDFLOWERS** too.

MAMMALS rely on areas of wildflowers as places to live and find food.

Wildflowers give **INSECTS** a home too!

BIRDS nest in areas with wildflowers and eat insects, seeds, and berries.

WHAT CAN YOU DO TO HELP?

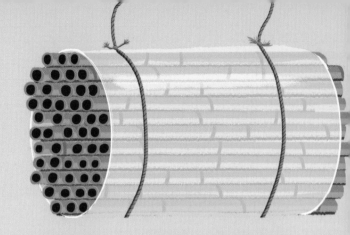

PLANT BEE-FRIENDLY PLANTS

The bees will thank you for planting flowers rich in nectar and pollen. Wildflowers are the most suitable for bees, but this doesn't mean you have to go planting weeds. Find out what to plant on the next page.

BEE CALM

Don't be scared of bees. If a bee flies near to you, then stay calm and move away slowly. It will soon realize you are not a flower and fly away.

BEE IN THE KNOW

Become a bee scientist by learning more about bees. Reading this book is a good start but there's a lot more to find out about different types of bees. Get outdoors and watch the bees at work. What will you discover?

SPREAD THE WORD

Bringing people together in your school or community could really help the bees. You could suggest a project to plant more bee-friendly plants in public spaces, or make bee hotels with your class.

MAKE A BEE HOTEL

Solitary bees like to nest in small holes. Give these bees a five-star place to stay by making a bee hotel.

YOU WILL NEED:

bamboo stems (enough to fill the inside of the bottle)

a 2 litre **plastic drink bottle**

twine

HERE'S HOW:

  Have you seen a **carpenter bee** visiting your hotel?

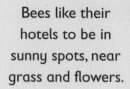

1.

Cut off both ends of the plastic bottle with the help of an adult.

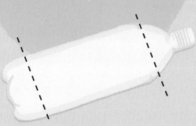

2.

Trim the bamboo sticks to the same length as the bottle (you'll need an adult's help again).

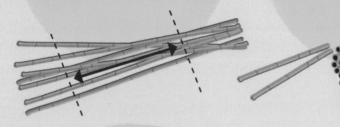

3.

Pack as much bamboo as will fit into the bottle. You could also fill it with sticks, pine cones, and bark.

Bees like their hotels to be in sunny spots, near grass and flowers.

4.

Tie twine tightly around both ends of the bottle.

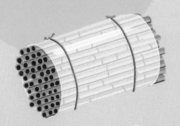

5.

Tie each end of a new piece of **twine** to the two twine loops. Make the string long enough for the bottle to hang.

6.

Hang the bee hotel on a fence or wall outdoors, 1m (3ft) above the ground. Make sure that the bee hotel is level and secure.

Keep checking on your bee hotel to see which different types of bees have checked in! They will be nesting in the hollow stems of the bamboo.

HAVE A BEE-FRIENDLY GARDEN

Bees love gardens that are full of flowers and different plants with plenty of nectar and pollen. Here are a few tips on how you can help bees by making some changes to your garden:

GET PEOPLE OUTDOORS

You might need some help with buying new plants and planting them in your garden, so ask family and friends to help. Everyone will enjoy getting outdoors.

PLANT LOTS OF FLOWERS

Bees like it when there are lots of different flowers with strong scents. If you don't have a garden, you could try planting in a window box or hanging basket.

LOVE THE WEEDS

Let your garden be a little on the wild side! Bees love wildflowers like dandelions and clover, so don't pull out all of the weeds and don't mow the grass too often.

AVOID USING CHEMICALS

Pesticides and other gardening products might promise to make your garden perfect, but they can be harmful to bees. You can still have a beautiful garden without using chemicals.

WHAT TO PLANT FOR THE SPRING:

CROCUS

SNOWDROP

IRIS

WHAT TO PLANT FOR THE SUMMER:

WHAT TO PLANT FOR THE AUTUMN:

LAVENDER

MINT

CORNFLOWER

SUNFLOWER

SEDUM

IVY

THE **BEE ALL** AND **END ALL**

It's sad to lose any animal, but to lose bees would be to lose nature's hardest worker. Bees are responsible for so much more than making honey. So let's appreciate our fuzzy friends by doing what we can to help them.

Keep bumbling on about the bees, because they really do matter.

ABOUT THE AUTHOR

Charlotte spends her time doing what she loves: creating books with playful designs to bring information to little readers. When she started researching bees, Charlotte became absorbed in their world of making honey and doing the waggle dance. She just had to make a book about it! Not only was it a good excuse to draw a bee in a tutu, but she wanted to share how bees are fundamental to the world we live in today. She hopes **The Bee Book** will spark a conversation between parent and child that inspires them to think about bees, make a bee hotel, or even plant a sunflower.

Charlotte studied illustration at Kingston University, where she discovered picture books as a tool to bring important topics to life. Since then, it has become her job, as she works with other book enthusiasts at DK publishing.

INDEX

Author and Illustrator Charlotte Milner
Editor Violet Peto
Senior Producer Amy Knight
Producer, Pre-Production Nikoleta Parasaki
Managing Editor Penny Smith
Managing Art Editor Mabel Chan
Art Director Jane Bull
Publisher Mary Ling

First published in Great Britain in 2018 by
Dorling Kindersley Limited
80 Strand, London, WC2R 0RL

A CIP catalogue record for this book
is available from the British Library.
ISBN: 978-0-2413-0518-8

Printed and bound in China

A WORLD OF IDEAS:
SEE ALL THERE IS TO KNOW

www.dk.com

ACKNOWLEDGEMENTS

With many thanks to tutors at Kingston University, in particular Jake Abrams, Mark Harris,
and Ben Newman who helped guide me through the early stages of the project. I would also
like to thank friends and family who have supported me whilst hearing about nothing but bees!